Written by OLAYINKA OGINNI
Illustrated by Tryandi MOCH
Edited by Scorpio & Claire Ormsby-Potter

Along with Mama B, one bright summers day,
Ola and Olu waited for a bus to come their way.

Olu pleaded to visit the market, for he had coins to spare!
"OK You can go," said Mama B. "But please take care!"

He promised Mama B., though he felt it a fuss,
to take Ola with him and rush back for the bus.

Arriving at the market, the pair were so happy!
"Here," said Olu to Ola, "choose some fruit, they're on me!"
Handing her some coins, he pointed towards the market stall.
"Would you rather an apple, orange, or pear, or perhaps them all?"

Looking around Olu saw more delights. So many good fruits, they could dine on that night!
He continued his questioning. "What about cherries, or strawberries, bananas,
or even blue berries!?"

"How many fruits may I take?" Ola wondered aloud.
"I can treat you to 10," Olu said rather proud.
Ola took the fruit one by one, beginning to count.
But as she did, her fears began to mount!
1,2,3. 1,2,3. 1,2,3........4?!
"Hurry up!" yelled Olu. "There's no time to get more!"

The bus was nearing the stop, you see,
and so, they both ran to rejoin Mama B.
They got on the bus, but hardly with ease!
For nearly missing it, Mama B. was not pleased!
Ola, upset, began to cry!
Olu felt bad, and questioned her why?

Olu gave Ola a hug, in a bid to be kind,
realising that Ola had left her precious fruits behind.

"I can't count well!" Ola cried in despair.
"I took too long to pick my fruit, so now I have none!"
"It's not fair!"

"Don't cry, now!" said Olu. ""I'll teach you to count!"
"Then next time, you'll be able, to select the right amount!"

1 2 3 4 5 6 7 8 9 10

So Olu began to teach Ola, to count to ten.
Promising they'd return to the market, to buy fruit once again!

1,2,3....... 1,2,3......4.
Harder and harder, Ola tried to count more!
1,2,3......4. 1,2,3,4......5.
Ola began to gain confidence. She felt so alive!

Though Ola had managed to conquer her fears,
and make it to 7, without any tears.
The joy she had felt, soon turned to hate,
as she struggled and struggled, to count up to 8!

1,2,3,4,5,6...... 1,2,3,4,5,6,7...... Ola tried to count higher.
To count up to 10, was her only desire.
"I can't do it!" in anger, Ola cried!
Olu smiled, "keep trying, you're sure to be surprised!"
4
3
5
2
6
1
7

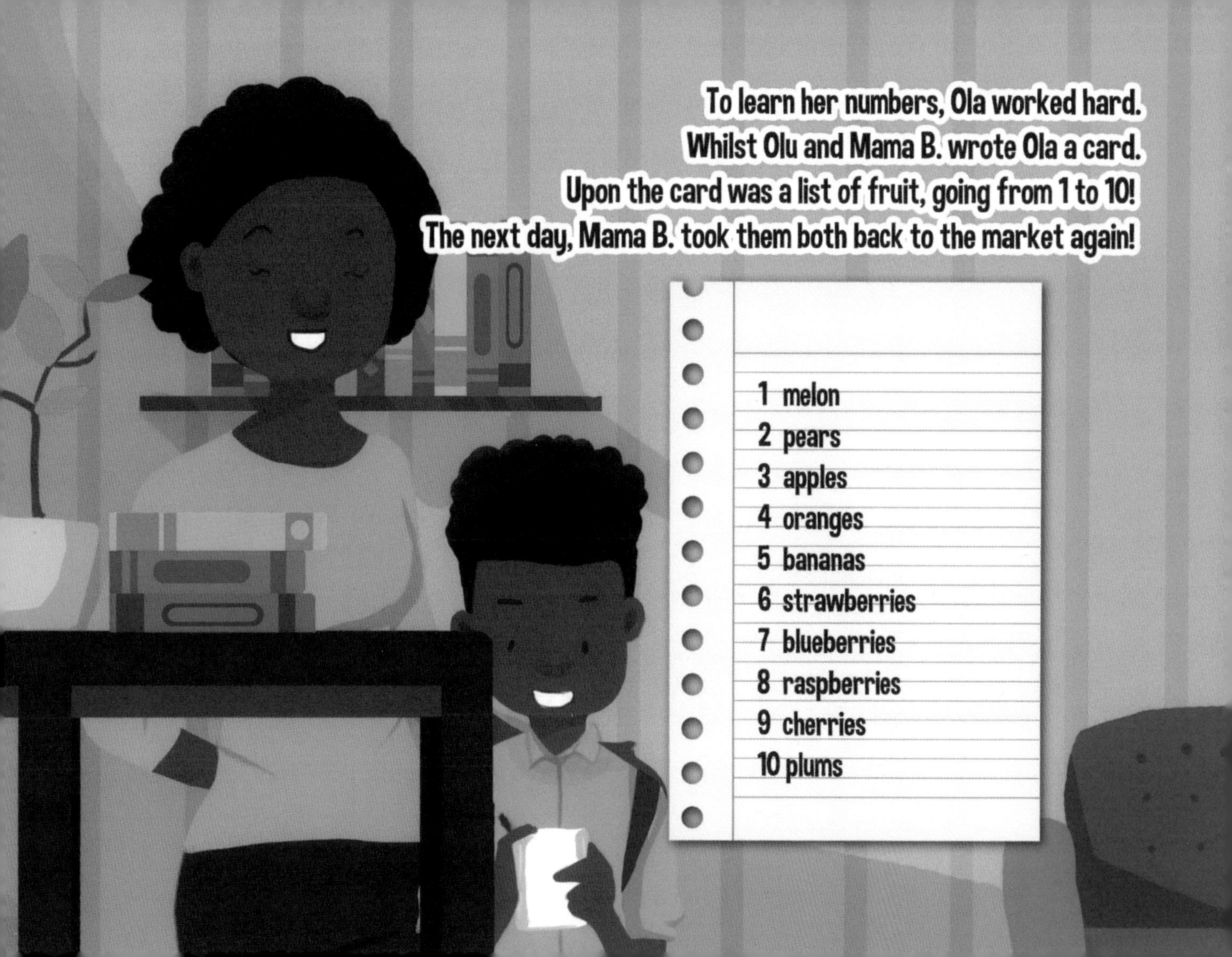
To learn her numbers, Ola worked hard.
Whilst Olu and Mama B. wrote Ola a card.
Upon the card was a list of fruit, going from 1 to 10!
The next day, Mama B. took them both back to the market again!
1 melon
2 pears
3 apples
4 oranges
5 bananas
6 strawberries
7 blueberries
8 raspberries
9 cherries
10 plums

She handed Olu the money and Ola the card,
and said, "you can do this Ola," "you've practiced so hard!"

So, under the watchful eyes of Mama B. and Olu,
Ola began to collect the fruit from the list,
reciting the numbers she knew!

She picked:
1 melon
2 pears
3 apples
4 oranges
5 bananas

1

2

3

4

5

6 7 8 9 10

6 strawberries
7 blueberries
8 raspberries
9 cherries
10 plums

1,2,3,4,5,6,7,8,9,10..........10!.........10.........10!
Ola couldn't believe it! She counted her fruit again and again!
"Well done," said Mama B. "I'm so proud, you got it right!"

And so, both Ola and Olu feasted on fresh fruit, later that night!

The

END

Printed in Great Britain
by Amazon

58372487R00015